UNUSUAL FARM ANIMALS

RAISING REINDEER

SARAH MACHAJEWSKI

Jefferson Twp. Public Library
1031 Weldon Road
Oak Ridge, NJ 07438
973-208-6244
www.jeffersonlibrary.net

PowerKiDS press
New York

Published in 2020 by The Rosen Publishing Group, Inc.
29 East 21st Street, New York, NY 10010

Copyright © 2020 by The Rosen Publishing Group, Inc.

All rights reserved. No part of this book may be reproduced in any form without permission in writing from the publisher, except by a reviewer.

First Edition

Editor: Tanya Dellaccio
Book Design: Michael Flynn

Photo Credits: Cover (reindeer) Bartosz Luczak/Shutterstock.com; (series barn wood background) PASAKORN RANGSIYANONT/Shutterstock.com; (series wood frame) robert_s/Shutterstock.com; cover, pp. 1, 3, 23, 24 (reindeer icon) draco77vector/Shutterstock.com; p. 5 Little Adventures/Shutterstock.com; p. 6 Sylvie Bouchard/Shutterstock.com; p. 7 Andrei Kobylko/Shutterstock.com; p. 9 longtaildog/Shutterstock.com; p. 11 Suwipat Lorsiripaiboon/Shutterstock.com; p. 12 NataSnow/Shutterstock.com; p. 13 Francisco V Machado/Shutterstock.com; p. 14 Edoma/Shutterstock.com; p. 15 VW Pics/Universal Images Group/Getty Images; p. 17 Marc Scharping/Shutterstock.com; p. 18 Standa Riha/Shutterstock.com; p. 19 BlueOrange Studio/Shutterstock.com; p. 21 Sergey Krasnoshchokov/Shutterstock.com; p. 22 Iakov Filimonov/Shutterstock.com.

Cataloging-in-Publication Data

Names: Machajewski, Sarah.
Title: Raising reindeer / Sarah Machajewski.
Description: New York : PowerKids Press, 2020. | Series: Unusual farm animals | Includes glossary and index.
Identifiers: ISBN 9781725309104 (pbk.) | ISBN 9781725309128 (library bound) | ISBN 9781725309111 (6 pack)
Subjects: LCSH: Reindeer–Juvenile literature.
Classification: LCC QL737.U55 M325 2020 | DDC 599.65'8–dc23

Manufactured in the United States of America

CPSIA Compliance Information: Batch #CWPK20. For Further Information contact Rosen Publishing, New York, New York at 1-800-237-9932.

CONTENTS

NOT JUST UP NORTH 4
WHAT ARE LIVESTOCK? 6
BUILT FOR THE COLD 8
WHY RAISE REINDEER? 10
REINDEER FARMING BASICS 14
FARMING CHALLENGES 16
THE REINDEER DIET 18
THE FUTURE OF REINDEER FARMING 20
GENTLE BEASTS 22
GLOSSARY 23
INDEX 24
WEBSITES 24

NOT JUST UP NORTH

When you think of reindeer, do you picture them at the snowy North Pole? Or do you picture them on the farm alongside pigs and chickens? Believe it or not, reindeer are farm animals! Many farmers love to raise them because they're friendly and gentle.

Farmers raise reindeer to **breed** and sell, to use for meat, and to enjoy as pets. Raising reindeer requires special **techniques** and important knowledge of how to care for them.

HOW UNUSUAL!

In some areas of the world, reindeer are called caribou.

REINDEER ARE COMMON ON FARMS ALL AROUND THE WORLD.

WHAT ARE LIVESTOCK?

"Livestock" is another word for farm animals. Farmers raise cows, sheep, and pigs as livestock. In some parts of the world, farmers might raise bison, oxen, or even camels!

HOW UNUSUAL!

Farmers in Russia and Scandinavia have raised reindeer as livestock for thousands of years.

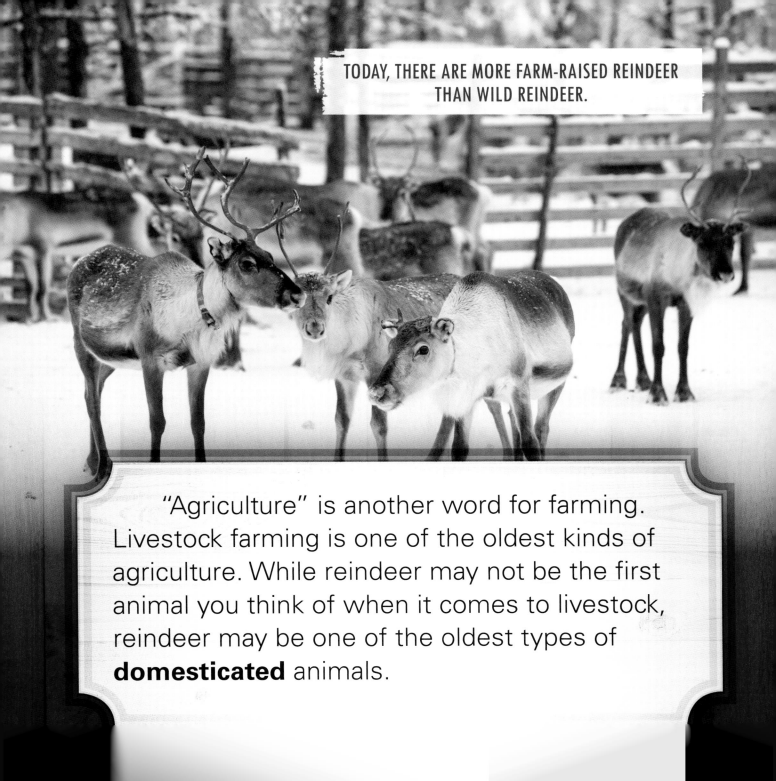

TODAY, THERE ARE MORE FARM-RAISED REINDEER THAN WILD REINDEER.

"Agriculture" is another word for farming. Livestock farming is one of the oldest kinds of agriculture. While reindeer may not be the first animal you think of when it comes to livestock, reindeer may be one of the oldest types of **domesticated** animals.

BUILT FOR THE COLD

Reindeer are well suited for cold weather. Their natural habitats, or homes, are in the snowy, wooded forests of Alaska, Canada, northern Asia, and northern Europe.

These animals are built to survive in cold areas of the world. Tall and heavy, reindeer can weigh between 240 to 700 pounds (108.9 to 317.5 kg). Their flat, wide hooves help them walk on snow and ice. They have two **layers** of hair to keep them warm.

HOW UNUSUAL!

Reindeer hair floats! Each hair in a reindeer's top coat is hollow, or empty in the middle.

BOTH MALE AND FEMALE REINDEER HAVE ANTLERS, OR BONY GROWTHS ON THEIR HEAD. THESE ANTLERS FALL OFF AND GROW BACK EVERY YEAR.

WHY RAISE REINDEER?

Reindeer are an important farm animal because they can be used for a lot of different purposes. In the past, some farmers used all parts of the animal to survive. They drank reindeer milk and used their skin and fur for clothing. They used antlers to make tools. Reindeer meat was an important food.

Today, reindeer are sometimes raised as pets. A reindeer's gentle nature can make it a good friend on the farm!

HOW UNUSUAL!

Reindeer sinew is what connects their muscles to their bones. People can use it to make sleds!

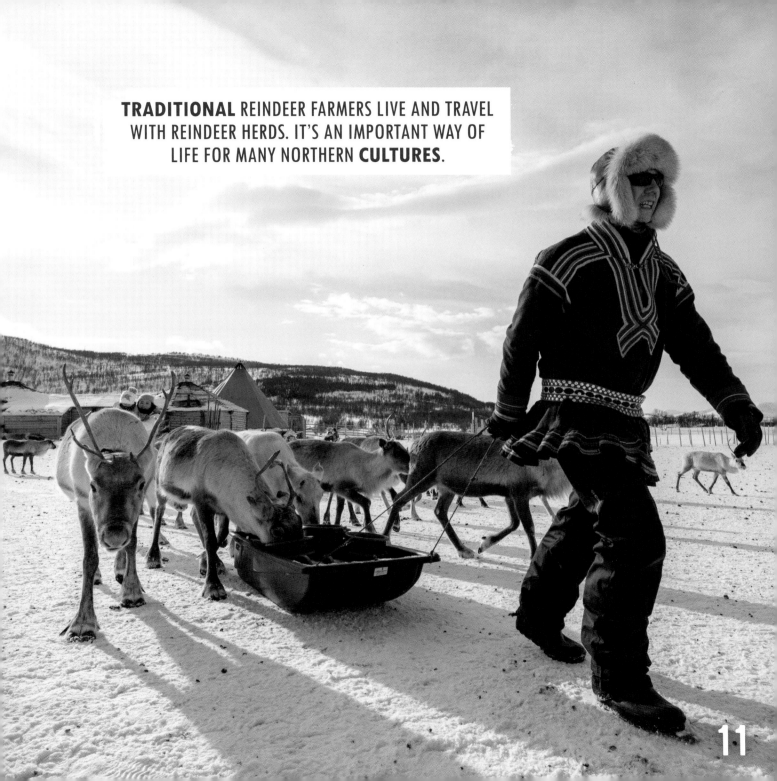

TRADITIONAL REINDEER FARMERS LIVE AND TRAVEL WITH REINDEER HERDS. IT'S AN IMPORTANT WAY OF LIFE FOR MANY NORTHERN **CULTURES**.

There are plenty of reasons to raise reindeer. Many people like to see reindeer during the winter months, around the holidays. Farmers who raise reindeer can make money by showing their reindeer at events. Additionally, reindeer can be trained to give people rides and to pull sleds.

MANY PEOPLE THINK OF REINDEER WHEN THEY THINK OF WINTER HOLIDAYS SUCH AS CHRISTMAS.

People also raise reindeer for their meat. Reindeer meat has less fat than many other kinds of livestock meat, so it's becoming more popular to cook with and eat.

REINDEER FARMING BASICS

Like humans, reindeer need food, water, and care to stay healthy. On most farms, reindeer live in fenced-in pens or barns. Some farmers let their reindeer **roam** freely.

Raising reindeer can also include training them. Many farmers train their animals to be comfortable around people. The animals learn commands such as "walk," "whoa," and "stand." They also may be trained to wear a **halter** and reins. This takes time and lots of treats!

PEOPLE HAVE BEEN RAISING REINDEER IN NORTHERN EUROPE FOR THOUSANDS OF YEARS. BELOW IS AN IMAGE OF A REINDEER FARM IN FINLAND.

FARMING CHALLENGES

As with any livestock, reindeer farming has challenges, or problems. Reindeer have thick fur coats. They can overheat, or get too hot, quickly. Farmers know they can't work their reindeer too hard when it's warm.

Another challenge is when reindeer are "in velvet," or growing antlers. During this time, their antlers are soft. Farmers need to handle their reindeer gently so they don't hurt them. Good farmers should learn how to deal with these challenges before they begin raising reindeer.

HOW UNUSUAL!

Sometimes free-roaming reindeer need to be rounded up. Farmers may do this on foot, on horseback, and even by helicopter.

REINDEER WITH NEW ANTLERS ARE CALLED "IN VELVET" BECAUSE THE SOFT SKIN THAT COVERS GROWING ANTLERS FEELS LIKE THE CLOTH CALLED VELVET.

THE REINDEER DIET

On the farm or in the wild, reindeer are grazing animals, which means they eat grass and other plants that grow where they roam. Wild reindeer eat grasses, moss, tree leaves, and other plants.

AN ADULT REINDEER CAN EAT BETWEEN 9 AND 18 POUNDS (4.1 TO 8.2 KG) OF FOOD A DAY.

On the farm, reindeer eat grass in their **pasture**. In addition, farmers give them hay, grain, and special biscuits and feed, or food, made just for reindeer. Farmers must remember to be sure their reindeer have plenty of food and water at all times.

THE FUTURE OF REINDEER FARMING

Reindeer herding has been a way of life for many cultures for thousands of years. However, today's herders have less land for their reindeer to roam. Additionally, fires and drought, or long periods of dry weather, are a problem for herders and their reindeer.

Throughout the United States and Canada, reindeer farming is a growing business. The call for reindeer at winter events keeps farmers in business. So does the need for reindeer meat. The future of reindeer farming looks bright!

EVEN THOUGH REINDEER FARMING IS A GROWING BUSINESS, THE REINDEER **POPULATION** IS FALLING. FARMERS AND HERDERS ARE WORKING TO KEEP THEIR LIVESTOCK STRONG.

GENTLE BEASTS

Demand for reindeer is growing, which is good news for farmers. Many farmers begin by adding a few reindeer to their farm. They care for them alongside other livestock. It doesn't cost a lot to keep and care for reindeer. These reasons and more make it easy for farmers to enjoy and profit, or make money, from raising reindeer. In addition, reindeer are great farm animals because they are gentle and calm. It's easy to see why farmers like to raise them.

GLOSSARY

breed: To bring a male and female animal together so they will have babies.

culture: The ways of life of a certain group of people

domesticated: Bred and raised for use by people.

halter: A rope or strap placed around an animal's head.

layer: One part of something lying over or under another

pasture: Land covered with grass or low plants.

population: The number of animals in a species that live in a place.

roam: To move about freely.

technique: A method of doing something.

traditional: Having to do with the ways of doing things in a culture that are passed down from parents to children.

INDEX

A
Alaska, 8
antlers, 9, 10, 16, 17
Asia, 8

C
Canada, 8, 20

E
Europe, 8, 15

F
Finland, 15
food, 14, 19
fur, 10, 16

M
meat, 4, 10, 13, 20

P
pets, 4, 10

R
Russia, 6

S
Scandinavia, 6

U
United States, 20

W
winter, 12, 20

WEBSITES

Due to the changing nature of Internet links, PowerKids Press has developed an online list of websites related to the subject of this book. This site is updated regularly. Please use this link to access the list: www.powerkidslinks.com/ufa/reindeer